THE SASQUATCH

(BIGFOOT) PEOPLE

JOIN ME IN MY
CRYPTOZOOLOGY EXPERIENCE

MICHAEL L. GLASGOW

TABLE OF CONTENTS

INTRODUCTION

Hello everyone, my name is Michael Glasgow, and this my account into Cryptozoology. I have always been, and will continue to be a religious person. I believe God has a plan for us all in this world. It just so happens that he has influenced me to be an open minded person. However, I am also a realist, and seeing is believing. The world we live in is a wondrous place. We have barely glimpsed what this world has to show us. So many wonderful people, and creatures that are here living in harmony with us. Many species of animals, and other beings on this planet we call Earth.

I will walk you through my Crypto experience with my own pictures taken by me, and me only. Nothing Doctored, or CGI. My pictures are some of the best out there, I believe. I am only leading the horse to the water, it's up to him to drink it in, or not. All I ask is that you have an open mind. Prepare yourself for a hike into the wild with me. Be couscous, but not afraid. I only offer the truth as you will see.

Chapter 1:

Do Not Awaken the Sleeping Giant

Entering the forest of Texas. I will not inform you of the exact location of my hairy friends for their protection, and yours. The Sasquatch (Bigfoot) people as I like to call them are in every forest of the world. They are an elusive people with the uncanny ability to camouflage themselves so well they could be right in front of you, and you could not see them. They are also a very curious, and watch us as we are in the forest. I'm sure you have probably heard many stories of Bigfoot looking in homes through the window. They like to watch, and interact with people from time to time. It is their curious nature in most cases. Think of them as another race of humans just hairier and much, much bigger. As in the case with most humans the majority of us are good law abiding citizens. We live to love our families, and live in harmony with most of the rest of the world making our meager life going about our day to day business, work, hobbies, and entertainment. We also have in our own kind that unfortunately are for lack of a better word just plain evil. Violent murders, thieving, lying, bush whacking scoundrels.

We as a people have good, and bad in our people. Keep in mind that the Sasquatch people are much like us. They also have good, and bad within their race as well. I believe that just like us they live in family units much like we do. There's a father, and mother with children. The father, and mother provide for the young just as we do. They have love for each other, and I'm sure it's much like ours. However you never know when you could run into a bad one that doesn't want humans around. If they get aggressive just leave the area.

When you visit the forest always be couscous for any animal. In most cases the animals, and the Sasquatch people will leave the area so they don't run into you. Although from time to time it does happen. Most encounters in many cases are harmless. A person may see a Sasquatch walking away from their direction. Sometimes you will get some playfulness like throwing small pebbles in your direction landing very close to you, or even hitting you. There are different actions like trees being pushed over or large, very large rocks being thrown. If that's happening most likely it's a territorial display. That means they want you out of the area, and suggest you get the heck out of there walk briskly don't run.

A sleeping Sasquatch deep in the bush.

Chapter 2:

<div align="center">~~~</div>

JUST CHILLING OUT
IN THE BUSHES

I FIRST DECIDED to hike in the forest just to enjoy Mother Nature's beauty as I have always enjoyed the outdoors. I went into the forest with an open mind believing in the possibility of Sasquatch. I like many people have watched the documentaries of Bigfoot. The foot castings and tree structures. As I was hiking I had found these odd tree structures. As I noticed them being woven, and inter locked into each other. It was clear to see that they didn't just grow like that someone had to have done it. It was no man that did it. Some of the trees were so large something with a great amount of strength had to have done it. They were bent over, and locked into another tree. They could possibly be some kind of shelter. Other tree breaks are believed by many Crypto zoologists that they are territorial markers meaning for others to keep out of the area.

Tree structure.

As I hiked along enjoying the forest listening for any sounds out of the ordinary. I would do some tree knocks here and there. They (Crypto zoologists) say it's a way of how they (Sasquatch people) communicate over long distances without calling out. I never received a response. I took pictures as I hiked along not really seeing anything in real time.

Sasquatch chilling in the bushes.

Chapter 3:

SASQUATCH IS WATCHING ME

Sasquatch friend is watching me from the bushes. I call him One Eyed Willy.

Another Big guy peeking through bushes.

WHILE I WAS hiking through the forest I would always check behind me. Listening to the forest noises for anything strange. I would take random pictures of the woods, and brush. Later I would review my pictures, low, and behold as I zoom in on dark spots. I found them in the bushes. There they were just poking their heads out watching me. They tend to be very still as to not attract attention to themselves. Sometimes they are hard to see, but as I zoom in you can see them more clearly. At first I could not believe my eyes. Then the more pictures that I took the more I could see. I thought to myself "Wow" they really are real, they are here, and they are watching me every time I enter the forest. I was amazed at sheer number that I found. I found many different ones even a small toddler was watching me. What's really funny was that I was feet away, and didn't even see him.

This is a little guy checking me out.
I'm guessing a toddler.

Chapter 4:

EATING TURTLES

WHEN I FOUND the first tree structure I found quite a few large turtle shells ripped open. The Sasquatch people must like to eat turtle. I noticed that they all were ripped open the same way which was one clue that made me think the Sasquatch people were eating them. The part of the turtle that was ripped open was the underside of the shell where the turtles head tucks back in when in defense mode. Something ripped it right off. As I examined the shell closely looking for scratch marks from teeth thinking to myself what animal did this. No coyote, or raccoon, possum could do this without leaving scratch marks from teeth. The only thing that could do this is something with great hand strength could do this, and leave no marks on the shell. Turtle is clearly part of Sasquatch people diet. I found several near each other not far from a tree structure. All were ripped open in the same manner.

Turtle shell ripped open.

Another shell

Another shell ripped open.

Yet another all within 15 foot radius.

Chapter 5:

~~~

Sasquatch Playing with My Wife

WHILE HIKING WITH my family on the day when we came across the turtle shells. I was walking about 30 yards in front of my wife and son. She yelled out, who is throwing pebbles at me?? She was hit in the face by a small pebble she said. I was a good distance away, and she knew it wasn't me. She said she was watching me when it happened. She knows it wasn't my son because he was knelling down next to her tying his shoe. I asked her are you sure it wasn't an insect that flew into her. She replied no, because when it hit my face I watched it drop to the ground. I explained to her that many Crypto researchers believe when they throw small pebbles they are being playful with you. Possibly they like you. She shrugged it off at the time just being unexplained. I later reviewed my pictures, and I found her friend watching us from behind a log about twenty yards away from us.

This is the juvenil that hit my wife with a pebble I'm sure of it.

I showed her the picture, and she was stunned in disbelief. She was a nonbeliever. Although she is still not convinced it reminds me of an old movie line. When they say "You're not going believe until it swims up and bites you on your @$$".

Now that I have seen with my own eyes I am a total believer. Over the years I have heard many stories that I must now believe there is truth to them.

Chapter 6:

⌇⌇⌇

NATIVE AMERICAN ACCOUNTS

THE NATIVE AMERICAN tribes all have names for the Sasquatch people. Sasquatch is just one of many names that the many tribes have for the same creature known to most of us as Bigfoot. Many tribes have learned to live with them. Many believe that if you leave them alone they will leave you alone. There are other stories by the Native American peoples. Some not so good. Many are stories of a rogue Sasquatch abducting a woman or child from the village, or while they were out gathering berries or water. It's possible that they are attracted to, or curious about the beautiful females. Possibly they wanted them as a pet or something to keep them company who really knows.

I think the Sasquatch people are much like humans at least in the way that most people are good, but just like humans there are bad ones as well. There was another story of a Caucasian man that was abducted by a large male Sasquatch while he was sleeping in his sleeping bag. He pushed him down in his bag and took him bag and all. The Sasquatch took him to a cave where he lived with his family. It was the large male father, mother and two young possibly teenagers. He had said it was a young male, and female but they were still taller than the man. They did not hurt the man but kept him like a pet possibly. He interacted with them for some time. He had said that the father Sasquatch liked his tobacco. The children were very curious as well. He ended up escaping to tell his story. I wonder what they really wanted from him. I would hope it wasn't anything harmful.

#13 Another young juvenile in the bush. I think this one may be a female.

A juvenile Sasquatch in the bushes.
Apparently they have white faces when they are young.

Chapter 7:

───~~───

THEY REVEALED THEMSELVES
TO ME

I HAVE HAD a feeling for a long time that they were there. They have the ability to make themselves invisible to us. They can make themselves invisible at will. If you can see them it's because they want you too, or they don't care that you do. Our eyes cannot see them in real time although they are there, but the camera can pick them up even when they are cloaked you can see the silhouette of a figure or a head much like a spirit, but they are not in spirit form. The fellow in this picture revealed himself to me just for a second. I was looking into the forest with a feeling that someone was there but I couldn't see him. I started taking pictures and then I felt the shiver go up my spine. You know the one like someone is watching just for a second. Then he de cloaked to show himself to me for a second. It was totally weird, but Awesome at the same time. I believe he hit me with infrasound to catch my attention as I was looking all over the forest. It was as if he hit me with the infrasound to get my attention, and say "Hey I'm over here" and there he was for about two seconds.

I believe he used infrasound on me because everyone knows if its adrenaline it lasts a lot longer than a second. You feel the fight or flight adrenaline feeling for quite a few minutes when that happens. We all know it that scared feeling, hyped up feeling with your heart beating out of your chest. I didn't feel that at all it was just like a zap, here I am. Then it was gone, and there he was looking at me. I waved to him, and then I proceeded back to my car. That was all I needed to see at that moment.

There he is right in the middle look for the silhouette.
Kind of looks Ghostly but believe me they are not.
They are all over the forest as many of them as we are in the cities.

Here are more Sasquatch people in the bushes trying to hide from us.
I have circled them.

Same picture without the circles can you spot them.
If you look closely here both the same picture I just circled them in one. There are a
few right in the center is a black one. Let me blow up this pic for you.

Close up head shot of the previous picture.

Here's a closer look at one in center.

Chapter 8:

THE MORE I LOOK
THE MORE I FIND

I WENT FISHING with my son on a beautiful spring day. Enjoying the quality time with my son at the Lake. I took pictures of all the surrounding area. We were at a spot with nice lagoon. Under a bridge was very nice except for the trash some fishermen left. I dropped my line in and while I was waiting for a bite. I began to pick up some of the trash. I only had a small plastic grocery bag, but it helped. I took pictures of opposite side of the lagoon. I captured some shots of the Sasquatch people watching us fish. Unfortunately I only caught one good size catfish. I would have shared if we could have got plenty. Sorry Big Guys next time.

Here is picture of one by the Lake.

This One is by the Lake too but reminds me character in movie. Sifi

Here's a picture of a family group. I think even a Dogman pup.

Here is picture of the other side of the lagoon we were fishing. Can't see anything from here, but as I zoom in the Big Guys were watching us from the other side.

I know this one is hard to see from across lagoon. Couple of them in shadows. Big guy!
I see his head part of his arm and hand resting on rock.

Chapter 9:

THE DOGMEN

WELL I WASN'T really planning on bringing the Dogman up, but since I put some pictures in the book with them I feel I should say something, and not leave them out.

The Dogmen are like the Sasquatch people except that they have heads of Dogs. They vary in size and look, but most of them are black. Most of the stories you hear about them are bad. It seems to me that, that can't be the case. I have been in the forest several times, and I can only get pictures of them. Never been accosted by Sasquatch people or Dogmen, and from my pictures I now I was close to them just yards away. I believe they are described to us from Ancient Egypt as the form of Anubis. I have pictures of them hanging out with the Sasquatch people in the bushes. They clearly have a symbiotic relationship of some kind.

Here is a picture of what I believe is a juvenile Dogman kicking back in the bush.
He looks like he's snarling. I can see his teeth. Hope it wasn't meant for me. They look
happy just relaxing in the in the forest. If you stop to think about it, it must be a won-
derful life. I would hope they are just as caring as the Sasquatch people.
However, I guess I am not one hundred percent sure about the Sasquatch people
I have much to learn.

THE SASQUATCH (BIGFOOT) PEOPLE

This little guy is a pup, but of course I'm not sure how old he is. I estimated him to about two and half to three feet tall. You can see he's definitely standing there with his arms to the side, and you can see hand not paw. Cute little guy. It's possible these are the descendants of Anubis here in modern day, but living in the forests.

Here is picture of a Dogman peeking out from behind a tree at me. I estimated him to be about five or six foot tall. Many stories about them of evil carnivorous beasts. I don't know how true it is I do take a firearm with me most of the time when I go out there. I catch them here and there on my camera. I know I have been pretty close to them, and I have seen no aggression towards me. They were within yards of me, and nothing happened except I got some good pictures.

This one is a little harder to see, but we have a Sasquatch person and a Dogman here. The two yellowish eyes are the Dogman.

I call this camocloaking. Look how green he is.

This guy was looking at me kind of hard I think. I hope he wasn't upset with me. I heard no noise so I guess he was just watching me.

Here is another guy lying deep in the bushes. Appears to be lying on him stomach peering at me through the bushes. No noticeable activity other than watching me.

THE SASQUATCH (BIGFOOT) PEOPLE

Here is a pretty big guy.

Well my friends, here is the last one for this adventure and he looks to be a BIG guy. It's been a lot of fun, I hope to have you back again. All I ask you is to have an open mind. Remember one thing, and that's that they are not the beasts you think they are. I advocate not hunting them, please know that they are a people. I wish you all much love, peace, and harmony

The End for now MG

www.ingramcontent.com/pod-product-compliance
Lightning Source LLC
Chambersburg PA
CBHW040110180526
45172CB00009B/1291